Instructor's Manual to Accompany

Clinical Calculations Made Easy

SOLVING PROBLEMS USING DIMENSIONAL ANALYSIS

SECOND EDITION

Gloria P. Craig, RN, MSN, EdD
Department Head for Nursing Student Services
and Assistant Professor
South Dakota State University
College of Nursing
Brookings, South Dakota

Lippincott
Philadelphia · New York · Baltimore

Ancillary Editor: Doris S. Wray
Project Editor: Debra Schiff
Senior Production Manager: Helen Ewan
Senior Production Coordinator: Michael Carcel
Art Director: Carolyn O'Brien
Manufacturing Manager: William Alberti

Second edition

9 8 7 6 5 4 3 2 1

ISBN: 0-7817-3241-7

■ Introduction

HOW TO USE THIS MANUAL

There are many reasons why learners have difficulty calculating medication problems including lack of mathematical abilities, mathematical anxiety, inability to conceptualize problems, and different learning styles. Teachers often contribute to the problem with inconsistencies in teaching methodologies and the use of different formulas that have a negative effect on the mathematical calculation abilities of the learners. Research suggests that learners have difficulty when multiple mathematical methods are used to teach medication calculation.

Dimensional analysis is a problem-solving method that can eliminate the stumbling blocks and empower learners with the ability to solve a variety of medication calculation problems using one problem-solving method. Dimensional analysis is a consistent problem-solving method that can be used to solve any type of medication calculation problem that the learner may come into contact with in clinical practice.

Dimensional analysis is based on cognitive theory and supports conceptualization of the medication calculation problem by offering the learner a consistent approach to setting up medication problems. Through critical thinking, the concept of dimensional analysis supports visualization of all parts of a medication calculation problem. Dimensional analysis uses a unit path grid that provides a neat and organized step-by-step flow of problem solving. The unit path grid consists of four terms that provide the basis for solving problems with dimensional analysis. These terms include

- *Given quantity* (the beginning point of the problem)
- *Wanted quantity* (the answer to the problem)
- *Unit path* (the series of conversions necessary to achieve the answer to the problem)
- *Conversion factors* (equivalents necessary to convert between systems of measurement and to allow unwanted units to be canceled from the problem)

Below is an example of the problem-solving method, showing the placement of the four basic terms used with dimensional analysis on the unit path grid.

Given quantity	Conversion factor for given quantity	Conversion factor for wanted quantity	Conversion computations		Wanted quantity
				=	

Once the given quantity is identified, the unit path leading to the wanted quantity is established. The problem-solving method of dimensional analysis can be explained using the following five steps:

1. Identify the given quantity in the problem.
2. Identify the wanted quantity in the problem.
3. Establish the unit path from the given quantity to the wanted quantity using equivalents as conversion factors.
4. Set up the problem to permit cancellation of unwanted units.
5. Multiply the numerators, multiply the denominators, and divide the product of the numerators by the product of the denominators to provide the numerical value of the wanted quantity.

Suggested use of overhead transparencies (OT) is indicated throughout this manual using the initials OT and a number in parentheses (eg, OT 3). The number relates to a specific transparency master that is located in the last section of this manual. The selection of OT masters will provide a solid foundation of visual aids to support the teaching of various aspects of dimensional analysis. Additional OTs can be developed to illustrate the many varieties of medication problems that can be solved using the five steps of dimensional analysis.

GENERAL TEACHING/LEARNING STRATEGIES

Current research suggests that several teaching/learning strategies can be implemented by the teacher and incorporated into the approach to teaching dimensional analysis that will maximize the learning process. Following are some of these strategies:

- Support student learning through providing approximately 40% to 50% of class time for student learning activities and in-class practice. Teaching/learning theories support the idea that learners retain 50% of what they see, 20% of what they hear, and 30% of what they do.
- Encourage the students to use their preferred learning style. Each student has a preferred learning style, which includes visual, auditory, kinesthetic-tactile, or a combination of the learning styles. Each chapter will provide examples of how the content can be presented using all three learning styles.
- Use music to stimulate the right side of the brain and assist the learner to relax. When a student is able to relax, assimilation and retention of information are increased. Music promotes relaxation and rapid assimilation of information. Promote relaxation through the use of classical background music during didactic presentations and baroque background music during the in-class practice time.
- Use color to stimulate the right side of the brain and assist the learner to increase alertness. When a student is more alert, visual recall of information is increased. Promote visual recall through the use of peripheral learning charts. These large and colorful charts can be placed on the walls around the classroom within the peripheral vision of the learner. Peripheral learning charts will provide the learner with a visual picture of the auditory information that they will need to know.
- Alleviate student "math phobia" using positive reinforcement and individual encouragement. Anxiety that produces math phobia is an issue that each learner must deal with and often requires small group work or individual tutoring and counseling.

CONCLUSION

After 15 years of teaching dimensional analysis, the best advice I can give to anyone who is going to teach dimensional analysis is: have FUN! Create a classroom environment that will encourage student participation, collaborative learning, and peer teaching. Medication calculation need not be difficult. You just need a problem-solving method that is easy to understand and implement. That method is dimensional analysis.

Contents

SECTION 1

CHAPTER 1

Arithmetic Review

INTRODUCTION

The first chapter focuses on basic arithmetic skills involving Arabic numbers and Roman numerals as well as multiplying and dividing fractions and decimals. It is very important to assess the ability of the learner to multiply and divide fractions and decimals because this provides the foundation for using dimensional analysis as a problem-solving method. It is also necessary to take into consideration the demographics of each class when teaching dimensional analysis. Each class will include various types of learners, multiple levels of mathematical abilities, and accompanying anxiety levels. For some learners, the anxiety level may be extremely high because they have not used basic arithmetic skills for many years, whereas other learners may have no difficulty demonstrating their ability to solve problems using basic arithmetic skills.

When teaching medication calculation, the instructor emphasizes that every nurse must know and practice the "five rights of medication administration." Although the right drug, route, time, and patient may be readily identified, the right DOSE requires mathematical, conceptual, and cognitive skills that may pose difficulty for some learners. Research has demonstrated that learners often have trouble with medication calculations because they cannot conceptualize the problem.

■ Objectives

This chapter assists the learner by reviewing basic mathematical skills that will prepare the learner to calculate medication dosage calculation problems using the problem-solving method of dimensional analysis. After completing this chapter, the learner will be able to

1. Express Arabic numbers as Roman numerals.
2. Express Roman numerals as Arabic numbers.
3. Identify the numerator and denominator in a fraction.
4. Multiply and divide fractions.
5. Multiply and divide decimals.
6. Convert fractions to decimals.

■ Teaching Tips

Some strategies used to stimulate the right side of the brain and to promote the learning of the content in this chapter include the use of

- Peripheral learning charts (large wall charts within the peripheral vision of the learner) placed around the room to provide visual examples of
  - Conversions between Roman numerals and Arabic numbers.
  - Definitions and examples of fractions clearly labeling the numerator, dividing line, and denominator.
  - Conversions of fractions to decimals.
- Abundant color in the peripheral learning charts. Color, a function of the right brain, assists the learner with rapid, visual recall and retention of material.

Some strategies to reduce mathematical anxiety that are particularly useful when teaching the content in this chapter include the use of

- Classical background music during the didactic presentation of the chapter concepts.
- Preassessment to identify the knowledge and skill level of each learner.
- Sample opportunity for small group work.
- Availability for individual assistance for each learner from either the teacher or student tutors.

■ Student Learning Activities for Major Concepts

CONCEPT 1. ANXIETY

- Invite the learners to share positive mathematical learning experiences.
- Invite the learners to share negative mathematical learning experiences.
- Discuss with the learners the different types of anxiety and how anxiety affects mathematical performance.

- Discuss with the learners coping strategies for dealing with "math phobia."
- Discuss preferred learning styles to promote successful learning.

CONCEPT 2. ARABIC NUMBERS AND ROMAN NUMERALS

- Discuss with the learners how Roman numerals are used in medication orders written by physicians.
- Discuss with the learners which Roman numerals are used most frequently in medication orders written by physicians.
- Give an example of a medication order written by a physician that contains a Roman numeral.

CONCEPT 3. FRACTIONS

- Discuss with the learners how fractions are used in medication orders written by physicians.
- Discuss with the learners which fractions are used most frequently in medication orders written by physicians.
- Give an example of a medication order written by a physician that contains a fraction.

CONCEPT 4. DECIMALS

- Discuss with the learners how decimals are used in medication orders written by physicians.
- Discuss with the learners which decimals are used most frequently in medication orders written by physicians.
- Give an example of a medication order written by a physician that contains a decimal.

■ Frequently Asked Questions

The most frequently asked questions regarding the information in Chapter 1: Arithmetic Review include

1. How often will medication orders contain Roman numerals?

   Answer: Medication orders seldom contain Roman numerals higher than 20, but Roman numerals that may be used include II (2), III (3), and X (10).

2. Which fraction do you invert when you are dividing fractions?

   Answer: Invert (turn upside down) the divisor portion of the problem (the divisor portion of the problem is the second fraction in the problem).

3. Can I convert fractions to decimals in every problem so that I don't have to work with fractions?

   Answer: Fractions can be converted to decimals for every problem if you prefer not to work with fractions. To convert a fraction to a decimal, divide the numerator portion of the fraction by the denominator portion of the fraction.

■ Conclusion

This chapter has reviewed basic arithmetic skills that will assist the learner to successfully implement dimensional analysis as a problem-solving method for medication dosage calculations. Regardless of the amount of time it takes, each learner should have a clear understanding of basic arithmetic skills and successfully complete the Practice Problems at the end of the chapter before proceeding to Chapter 2 and the concept of dimensional analysis.

POST-TEST FOR CHAPTER 1: ARITHMETIC REVIEW

Name ______________________ Date __________

Converting Between Arabic Numbers and Roman Numerals

1. 4 = _____

2. IX = _____

Answers:

1. IV

2. 9

Multiplying and Dividing Fractions

3. $\frac{1}{8} \times \frac{1}{8} =$ _____

4. $\frac{2}{4} \times \frac{1}{2} =$ _____

5. $\frac{1}{6} \div \frac{1}{3} =$ _____

6. $\frac{3}{4} \div \frac{7}{8} =$ _____

Answers:

3. $\frac{1 \times 1 = 2\ (2) = 1}{8 \times 8 = 64\ (2) = 32}$

4. $\frac{2 \times 1 = 2\ (2) = 1}{4 \times 2 = 8\ (2) = 4}$

5. $\frac{1 \div 1 = 1 \times 3 = 3\ (3) = 1}{6 \div 3 = 6 \times 1 = 6\ (3) = 2}$

6. $\frac{3 \div 7 = 3 \times 8 = 24\ (4) = 6}{4 \div 8 = 4 \times 7 = 28\ (4) = 7}$

(Post test continues on page 4)

Converting Fractions to Decimals

7. $\frac{1}{2} =$ ____

8. $\frac{3}{4} =$ ____

Answers:

7. $\frac{1}{2} = 0.5$

$$\begin{array}{r} 0.5 \\ 2\overline{)1.0} \\ \underline{1\,0} \end{array}$$

8. $\frac{3}{4} = 0.75$

$$\begin{array}{r} 0.75 \\ 4\overline{)3.00} \\ \underline{2\,8} \\ 20 \\ \underline{20} \end{array}$$

Multiplying and Dividing Decimals

9. 0.25 × 1.25 = _____

10. 0.125 ÷ 0.25 = _____

Answers:

9. 0.25 (2 decimal points)
×1.25 (2 decimal points)
0125
0500
02500
0.3125 (4 decimal points from right to left)

10. $0.25\overline{)0.125}$

(Move decimal points 2 places to the right)

Answer: 0.5

$$\begin{array}{r} 0.5 \\ 0.25\overline{)012.5} \\ \underline{125} \end{array}$$

CHAPTER 2

Systems of Measurement and Common Equivalents

INTRODUCTION

The second chapter assists the learner to understand the systems of measurement used for medication dosage administration. It is necessary for the learner to understand completely the systems of measurement in order to accurately implement the problem-solving method of dimensional analysis. To be able to calculate medication problems, the learner must have an understanding of common equivalents so he or she can visualize all parts of a medication dosage calculation problem. Visualizing all parts of the medication problem using dimensional analysis allows conceptualization of a medication dosage calculation problem. Through critical thinking, the concept of dimensional analysis supports visualization of all parts of a problem.

■ Objectives

This chapter assists the learner to understand the systems of measurement used for medication dosage calculation. After completing this chapter, the learner will be able to

1. Identify measurements included in the metric, apothecary, and household systems.
2. Understand abbreviations used in the metric, apothecary, and household systems.

■ Teaching Tips

This chapter will assist the learner to conceptualize through visualization. Students retain 50% of what they see; therefore, it is extremely important to have visual aids when teaching the student about common equivalents.

Stimulation of the right side of the brain is accomplished through the use of peripheral learning charts (Figures 2.1, 2.2, and 2.3) and by using different colors for the three different systems of measurement. These charts may be placed on the walls around the classroom within the peripheral vision of the learners. Visualizing the differences between the systems of measurement is an important part of conceptualizing any medication problem.

Hands-on experience with measuring devices and containers (medication cup and syringe; liter, gallon, quart, and pint bottles; and a glass, cup, tablespoon, and teaspoon) will assist the student to understand the differences between the systems of measurement. Students retain 30% of what they do, and hands-on experience provides the vehicle for understanding the systems of measurement. It is helpful for students to have a medication cup and medication syringe in front of them when solving medication problems to assist with conceptualization through visualization.

■ Student Learning Activities for Major Concepts

CONCEPT 1. MEASUREMENTS IN THE METRIC, APOTHECARY, AND HOUSEHOLD SYSTEMS

- Set up a weight station where learners can weigh-in and convert their weight from pounds to kilograms.
- Set up a weight station using a baby scale and a variety of dolls where learners can practice converting weight from kilograms to grams.
- Set up a measurement station where learners can practice using the measurement systems to measure colored water, flour, and dry Jell-O.
- Discuss with the learners how the three systems of measurement assist with accurately administering the Right Dose to the Right Patient.

CONCEPT 2. ABBREVIATIONS IN THE METRIC, APOTHECARY, AND HOUSEHOLD SYSTEMS

- Discuss with the learners how abbreviations are used in medication orders written by physicians.
- Discuss with the learners that most health care organizations have an approved list of abbreviations.

■ Frequently Asked Questions

The most frequently asked questions regarding the information in Chapter 2: Systems of Measurement and Common Equivalents include

1. Do I have to know all of the common equivalents?

   Answer: To promote safe administration of medication, the learner should have a clear understanding of all the common equivalents used in medication administration.

2. Do I have to memorize all of the common equivalents?

   Answer: It is not necessary to memorize all of the common equivalents because most nursing drug references contain a printed chart of common equivalents.

3. Can I laminate the conversion chart and carry it in my pocket?

   Answer: To promote safe administration of medication regardless of the clinical practice setting, a quick pocket reference would be helpful when a nursing drug reference is not available.

■ Conclusion

This chapter has reviewed the three systems of measurement that will assist the learner with being able to conceptualize the common equivalents necessary when problem solving with dimensional analysis. Each learner should have a clear understanding of the metric, apothecary, and household systems of measurement, be able to differentiate between the different abbreviations, and successfully complete the Practice Problems before proceeding to Chapter 3.

POST-TEST FOR CHAPTER 2: SYSTEMS OF MEASUREMENTS AND COMMON EQUIVALENTS

Name ______________________ Date ____________

1. 2.2 lb = _____ kg
2. 16 fl oz = _____ pt
3. 1 tsp = _____ mL
4. 15 gr = _____ g
5. 1 oz = _____ mL
6. 1000 mcg = _____ mg
7. 60 mg = _____ gr
8. 1 pt = _____ mL
9. 1 cc = _____ mL
10. 1000 mg = _____ g
11. 1 L = _____ mL
12. 4 qt = _____ gal
13. 1 tbsp = _____ tsp
14. 1 glass = _____ oz
15. 8 oz = _____ mL
16. 3 tsp = _____ mL
17. 15 gtt = _____ M
18. 1 dr = _____ cc
19. 15 M = _____ mL
20. 1000 g = _____ kg

(Post test continues on page 8)

Answers:

1. 2.2 lb = 1 kg
2. 16 fl oz = 1 pt
3. 1 tsp = 5 mL
4. 15 gr = 1 g
5. 1 oz = 30 mL
6. 1000 mcg = 1 mg
7. 60 mg = 1 gr
8. 1 pt = 500 mL
9. 1 cc = 1 mL
10. 1000 mg = 1 g
11. 1 L = 1000 mL
12. 4 qt = 1 gal
13. 1 tbsp = 3 tsp
14. 1 glass = 8 oz
15. 8 oz = 240 mL
16. 3 tsp = 15 mL
17. 15 gtt = 15 M
18. 1 dr = 5 cc
19. 15 M = 1 mL
20. 1000 g = 1 kg

CHAPTER 3

Solving Problems Using Dimensional Analysis

INTRODUCTION

Chapter 3 defines and explains the problem-solving method of dimensional analysis. Dimensional analysis provides a systematic, straightforward way to set up problems and helps to organize and evaluate data. This method has been used for many years in the field of chemistry because it is easy to learn and reduces errors when some type of mathematical conversion is required. Dimensional analysis inspires conceptualization and visualization of all parts of a problem. This chapter introduces dimensional analysis using a five-step approach. The learner is encouraged to practice solving problems involving common equivalents to enhance learning. "Thinking It Through" highlights insights into each part of the problem-solving method.

■ Objectives

After completing this chapter, the learner will be able to

1. Define the terms used in dimensional analysis.
2. Explain the step-by-step problem-solving method of dimensional analysis.
3. Solve problems involving common equivalents using dimensional analysis as a problem-solving method.

■ Teaching Tips

This chapter introduces the learner to new terms and concepts. It is important that the learner understand the four terms used with dimensional analysis (*given quantity, wanted quantity, unit path,* and *conversion factors*) and the five steps implemented when solving a problem using dimensional analysis.

- Use peripheral learning charts to demonstrate the unit path and encourage student participation in the learning process to facilitate retention of the information.
- Ask students to use the blackboard or whiteboard to demonstrate their problem-solving abilities. This approach actively involves the students with peer teaching while assisting the instructor to identify conceptual problems, areas that need clarification for the learner, and the need for additional in-class practice time. Because students only retain 20% of what they hear, lecturing over the material does not guarantee that the material has been understood, but active student participation demonstrates whether conceptualization has occurred for the learner.
- Provide in-class time for students to solve problems.

■ Student Learning Activities for Major Concepts

CONCEPT 1. DIMENSIONAL ANALYSIS

- Discuss with the learners the definition of dimensional analysis.
- Identify which learners have previous knowledge of dimensional analysis.

CONCEPT 2. THE FIVE STEPS OF DIMENSIONAL ANALYSIS

- Discuss with the learners the definitions of given quantity, wanted quantity, unit path, and conversion factors.
- Provide definitions (OT 1).
- Ask students to identify each of these in a problem (OT 4). As each answer is verbalized, write in the response in the appropriate space on the overhead transparency.
- Discuss with the learners the five steps of dimensional analysis (OT 2).
- Discuss with the learners how dimensional analysis allows conceptualization and visualization of all parts of the medication calculation problem. Illustrate the visualization aspect on overhead transparencies showing the five steps to take when solving a problem using the unit path grid (OT 3).

- Discuss with the learners how the terms *numerator* and *denominator* apply to dimensional analysis.
- Using the unit path grid (OT 8), demonstrate the concept of cancellation used with dimensional analysis by drawing a line through the unwanted units. Using another problem illustration on an overhead transparency, ask for student volunteers to repeat the demonstration.
- Demonstrate each step in dimensional analysis by solving problems (OT 4 through OT 9) and (OT 10 through OT 15).

■ Frequently Asked Questions

The most frequently asked questions regarding the information in Chapter 3: Solving Problems Using Dimensional Analysis include:

1. How do you know if the problem is set up correctly?

   Answer: If you are able to cancel all unwanted units from the unit path, then the problem is set up correctly. If you are unable to cancel unwanted units because both units are numerators or denominators, then the problem is not set up correctly. The units need to be opposite each other to be canceled from the problem.

2. Do you have to cancel the zeroes from the unit path?

   Answer: If you prefer, cancel only the unwanted units from the unit path. Other factors that can be canceled from the problem include like numerical values in the numerator and denominator portion of the problem and the same number of zeroes in the numerator and denominator portion of the problem.

3. Should I always circle the wanted quantity in the unit path?

   Answer: Circling the wanted quantity in the unit path identifies that the problem is set up correctly because the placement of the wanted quantity in the unit path correlates with the wanted quantity in the answer.

■ Conclusion

This chapter introduced the learner to the problem-solving method of dimensional analysis with a step-by-step explanation and an opportunity to practice solving problems involving common equivalents. This chapter has assisted the learner to

- Define dimensional analysis.
- Explain step-by-step the problem-solving method of dimensional analysis.
- Solve problems involving common equivalents using dimensional analysis as a problem-solving method.

Each learner should be able to successfully complete the Practice Problems at the end of the chapter before proceeding to Chapter 4.

POST-TEST FOR CHAPTER 3: SOLVING PROBLEMS USING DIMENSIONAL ANALYSIS

Name ______________________________ Date ______________

Changing units of measurement

1. 2045 g = How many lb?
2. 1/150 gr = How many mg?
3. 0.004 g = How many mcg?
4. 6 tsp = How many dr?
5. 0.5 L = How many pt?
6. How many L in 250 oz?
7. How many tbsp in 30 cc?
8. How many minims in 60 cc?
9. How many g in 45 gr?
10. How many oz in 1800 g?

Answers:

1.

2045 ~~g~~	~~1 kg~~	2.2 (lb)	2045 × 2.2	4499
	1000 ~~g~~	~~1 kg~~	1000	1000

= 4.499 or 4.5 lb

2.

$\frac{1}{150}$ ~~gr~~	~~1 g~~	1000 (mg)	$\frac{1 \times 1000}{150 \quad 1}$	$\frac{1000}{150}$	6.66
	15 ~~gr~~	~~1 g~~	15	15	15

= 0.44 or 0.4 mg

3.

0.004 ~~g~~	1000 ~~mg~~	1000 (mcg)	0.004 × 1000 × 1000	4000
	1 ~~g~~	1 ~~mg~~	1 × 1	1

= 4000 mcg

4.

6 ~~tsp~~	~~5 cc~~	~~1~~ (dr)	6
	~~1 tsp~~	~~5 cc~~	

= 6 dr

(Post test continues on page 12)

5. $\frac{0.5\ \cancel{L}}{} \Big| \frac{\cancel{1000\ mL}}{\cancel{1\ L}} \Big| \frac{\cancel{1\ qt}}{\cancel{1000\ mL}} \Big| \frac{2\ \text{(pt)}}{\cancel{1\ qt}} \Big| \frac{0.5 \times 2}{} \Big| \frac{1}{} = 1\ \text{pt}$

6. $\frac{25\cancel{0}\ \cancel{oz}}{} \Big| \frac{3\cancel{0}\ \cancel{mL}}{\cancel{1\ oz}} \Big| \frac{\cancel{1}\ \text{(L)}}{10\cancel{00}\ \cancel{mL}} \Big| \frac{25 \times 3}{10} \Big| \frac{75}{10} = 7.5\ \text{L}$

7. $\frac{30\ \cancel{cc}}{} \Big| \frac{1\ \cancel{tsp}}{5\ \cancel{cc}} \Big| \frac{1\ \text{(tbsp)}}{3\ \cancel{tsp}} \Big| \frac{30 \times 1 \times 1}{5 \times 3} \Big| \frac{30}{15} = 2\ \text{tbsp}$

8. $\frac{60\ \cancel{cc}}{} \Big| \frac{\cancel{1\ dram}}{5\ \cancel{cc}} \Big| \frac{60\ \text{(minims)}}{\cancel{1\ dram}} \Big| \frac{60 \times 60}{5} \Big| \frac{3600}{5} = 720\ \text{minims}$

9. $\frac{45\ \cancel{gr}}{} \Big| \frac{6\cancel{0}\ \cancel{mg}}{\cancel{1\ gr}} \Big| \frac{\cancel{1}\ \text{(g)}}{100\cancel{0}\ \cancel{mg}} \Big| \frac{45 \times 6}{100} \Big| \frac{270}{100} = 2.7\ \text{g}$

10. $\frac{18\cancel{00}\ \cancel{g}}{} \Big| \frac{\cancel{1\ kg}}{10\cancel{00}\ \cancel{g}} \Big| \frac{2.2\ \cancel{lb}}{\cancel{1\ kg}} \Big| \frac{16\ \text{(oz)}}{\cancel{1\ lb}} \Big| \frac{18 \times 2.2 \times 16}{10} \Big| \frac{633.6}{10} = 63.36 \text{ or } 63\ \text{oz}$

CHAPTER 4

One-Factor Medication Problems

INTRODUCTION

This chapter focuses on two major concepts: learning how to interpret medication orders correctly and how to calculate one-factor medication problems accurately using the five steps of dimensional analysis as a problem-solving method. To fully understand how to interpret medication orders and be able to calculate one-factor medication problems, the learner must have a clear understanding of the five rights of medication administration.

To be able to administer medication accurately, the five rights of medication administration form the foundation of communication between the physician and the nurse. The physician writes a medication order using elements of the five rights, and the nurse administers the medication to the patient based on the elements of the five rights. There may be a slight variation in the way each physician writes a medication order, but information pertaining to the five rights should be included in the medication order to ensure safe administration by the nurse. Both the physician and the nurse must adhere to the five rights to ensure that the right patient receives the right drug in the right dosage through the right route at the right time.

One of the essential components of the five rights is the right dose. To ensure that the patient receives the right dose, the learner must know how to read a medication label. This chapter provides the learner with the opportunity to experience hands-on opportunities to learn the components that form a drug label.

The five steps used in problem solving with dimensional analysis are introduced by explaining the four basic terms used in one-factor medication problems and the placement of the components of the medication order on the unit path. Using actual examples of medication orders and actual medication labels, the learner is guided step by step through solving one-factor medication problems with dimensional analysis.

This chapter also introduces the learner to the sequential method and the random method of solving problems using dimensional analysis. Each medication problem is explained and illustrated using both methods to provide the learner with an opportunity to practice and fully comprehend each method.

■ Objectives

After completing this chapter, the learner will be able to

1. Interpret medication orders correctly, based on the five rights of medication administration.
2. Identify components from a drug label that are needed for accurate medication administration.
3. Describe the different routes of medication administration: tablets and capsules, liquids given by medicine cup or syringe, and parenteral injections using different types of syringes.
4. Calculate medication problems accurately from the one-factor–given quantity to the one-factor–wanted quantity using the sequential or random method of dimensional analysis.

■ Teaching Tips

- It is important to actively involve the learner.
- Use peripheral learning charts with enlarged, colorful drug labels, medication cups, and a variety of syringes.
- Encourage hands-on experience using the drug labels provided in the textbook. Practicing with tablets and capsules, liquids given by medicine cup or syringe, and different types of medication syringes will assist the learner to conceptualize and assimilate the information.
- It is helpful to provide each learner with a kit containing a medication cup and syringe for liquid medications, medication vials or bottles, as well as the different types of parenteral syringes (3-cc syringe, low-dose and regular insulin syringes, and tuberculin syringe).
- The combination of calculating the problem (left brain) and hands-on experience (right brain) will assist the learner to implement the mathematical, conceptual, and cognitive skills necessary to calculate medication dosage problems using dimensional analysis.

■ Student Learning Activities for Major Concepts

CONCEPT 1. THE FIVE RIGHTS OF MEDICATION ADMINISTRATION

- Discuss the five rights of medication administration.
- Explore the meaning of each of the five rights.
- Discuss with the learners how physicians adhere to the five rights when writing prescriptions for patients.
- Discuss how nurses adhere to the five rights in practicing safe medication administration.
- Invite the learners to share situations in which they or a member of their family may have received a wrong drug or dosage of medication.

CONCEPT 2. ONE-FACTOR MEDICATION PROBLEMS

- Discuss with the learners how one-factor medication orders from physicians will contain only a numerator for the given quantity and the wanted quantity.
- Discuss with the learners the components of the medication order and the five steps of problem solving with dimensional analysis.
- Identify for the learners that the dosage of medication on hand is part of the unit path.
- Demonstrate for the learners the five steps of problem solving with dimensional analysis using overhead transparencies for each of the medication problem examples provided on pages 57 through 61 (OT 17 through 22).
- Explain to the learners the concept of the sequential method of dimensional analysis for each of the medication problem examples.
- Differentiate for the learners the sequential method and the random method of dimensional analysis using Example 4.3 on page 60 (OT 16).
- Set up learning stations for the medication orders involving aspirin, Advil, and Tylenol (pages 57 through 61) with the medication bottles and a physician order form for each station.

CONCEPT 3. COMPONENTS OF A DRUG LABEL

- Encourage the learners to identify the components of a drug label.
- Demonstrate examples of drug labels, and point out the components using overhead transparencies.
- Invite the learners to identify the components of the drug labels using overhead transparencies.
- Compare and contrast for the learners components of the drug label using capsules, tablets, oral liquids, injectable medications, intravenous medications, topicals and ointments, and rectal and vaginal suppositories.

CONCEPT 4. ROUTES OF MEDICATION ADMINISTRATION

- Discuss with the learners the different routes by which medication may be administered.
- Discuss with the learners the different devices used to administer medications.
- Set up a learning station for oral medication administration practice containing paper and plastic medication cups and medication syringes and bottles of colored water and bottles of M & Ms (or other kinds of tablet candies).
- Set up a learning station for parenteral medication administration practice containing a 3-cc syringe, low-dose and regular insulin syringes, a tuberculin syringe, and vials of colored water.

■ Frequently Asked Questions

The most frequently asked questions regarding the information in Chapter 4: One-Factor Medication Problems include:

1. Do all medication orders from physicians contain only one factor?

   Answer: Most medication orders obtained from a physician will contain one factor, but in Chapter 5 and Chapter 6 you will be introduced to two-factor and three-factor medication problems.

2. Is it better to use the sequential method or the random method to solve medication problems with dimensional analysis?

   Answer: Whether you use the sequential method or the random method of dimensional analysis, the important thing to remember is that dimensional analysis is a problem-solving method that uses critical thinking and is not a specific formula. Therefore, the important concept to remember is that all unwanted units must be canceled from the unit path and that the wanted quantity in the unit path correlates correctly with the wanted quantity in the answer.

3. When administering medication with a medication cup or syringe, do you measure below the line or right on the line?

   Answer: When administering medication with a

medication cup or syringe, the correct amount to pour or draw up is measured on the line of the medication cup or syringe.

4. Can I use another type of syringe to administer insulin?

   Answer: Insulin is administered specifically with an insulin syringe that requires no calculation. The number of units of insulin ordered by the physician equals the number of units drawn up in the insulin syringe. There are two types of insulin syringes: low-dose (each line on the syringe is a unit with a total volume of 0.5 cc or 50 Units) and regular (each line on the syringe is two units with a total volume of 1 cc or 100 Units).

■ Conclusion

This chapter has assisted the learner to interpret medication orders and drug labels correctly and calculate one-factor–given quantity to one-factor–wanted quantity medication problems accurately using the sequential or random problem-solving method of dimensional analysis. Each student should be able to complete the Practice Problems before proceeding to Chapter 5 and two-factor–given quantity to two-factor–wanted quantity medication problems.

POST-TEST FOR CHAPTER 4: ONE-FACTOR MEDICATION PROBLEMS

Name ______________________________ **Date** ______________

1. Order: Micronase 1.25 mg PO daily for non-insulin-dependent diabetes mellitus. How many tablets will you give?

   Answer:

   $$\frac{1.25\ \cancel{mg}}{\ } \Bigg| \frac{tablet}{2.5\ \cancel{mg}} \Bigg| \frac{1.25}{2.5} = 0.5\ tablet$$

2. Order: Tegretol 50 mg PO qid for seizures. How many tablets will you give?

   Answer:

   $$\frac{5\cancel{0}\ \cancel{mg}}{\ } \Bigg| \frac{tablet}{10\cancel{0}\ \cancel{mg}} \Bigg| \frac{5}{10} = 0.5\ tablet$$

3. Order: acetaminophen 240 mg PO every 4 hours prn for moderate pain. How many milliliters will you give?

   Answer:

   $$\frac{24\cancel{0}\ \cancel{mg}}{\ } \Bigg| \frac{5\ mL}{16\cancel{0}\ mg} \Bigg| \frac{24 \times 5}{16} \Bigg| \frac{120}{16} = 7.5\ mL$$

4. Order: lactulose 30 g PO qid for hepatic encephalopathy. How many milliliters will you give?

   Answer:

   $$\frac{3\cancel{0}\ \cancel{g}}{\ } \Bigg| \frac{15\ mL}{1\cancel{0}\ \cancel{g}} \Bigg| \frac{3 \times 15}{1} \Bigg| \frac{45}{1} = 45\ mL$$

5. Order: Tagamet 300 mg PO qid for short-term treatment of active ulcers. How many teaspoons will you give?

   Answer:

   $$\frac{\cancel{300\ mg}}{\ } \Bigg| \frac{\cancel{5\ mL}}{\cancel{300\ mg}} \Bigg| \frac{1\ tsp}{\cancel{5\ mL}} = 1\ tsp$$

(Post test continues on page 18)

6. Order: Tigan 0.2 g IM tid prn for nausea. How many milliliters will you give?

   Answer:

$$\frac{0.2\ \cancel{\text{g}}}{} \left| \frac{\text{mL}}{1\cancel{00}\ \cancel{\text{mg}}} \right| \frac{10\cancel{00}\ \cancel{\text{mg}}}{1\ \cancel{\text{g}}} \left| \frac{0.2 \times 10}{1 \times 1} \right| \frac{2}{1} = 2\ \text{mL}$$

7. Order: hydromorphone 3 mg IM every 3 hours for pain. How many milliliters will you give?

   Answer:

$$\frac{3\ \cancel{\text{mg}}}{} \left| \frac{\text{mL}}{2\ \cancel{\text{mg}}} \right| \frac{3}{2} = 1.5\ \text{mL}$$

8. Order: magnesium sulfate 1000 mg IM in each buttock for hypomagnesemia. How many milliliters will you give?

   Answer:

$$\frac{\cancel{1000}\ \cancel{\text{mg}}}{} \left| \frac{2\ \text{mL}}{\cancel{1}\ \cancel{\text{g}}} \right| \frac{\cancel{1}\ \cancel{\text{g}}}{\cancel{1000}\ \cancel{\text{mg}}} \left| \frac{2}{} \right. = 2\ \text{mL}$$

9. Order: naloxone HCl 200 mcg IV stat for respiratory depression. How many milliliters will you give?

   Answer:

$$\frac{2\cancel{00}\ \cancel{\text{mcg}}}{} \left| \frac{\text{mL}}{0.4\ \cancel{\text{mg}}} \right| \frac{1\ \cancel{\text{mg}}}{10\cancel{00}\ \cancel{\text{mcg}}} \left| \frac{2 \times 1}{0.4 \times 10} \right| \frac{2}{4} = 0.5\ \text{mL}$$

10. Order: Solu-Medrol 40 mg IM daily for autoimmune disorder. How many milliliters will you give?

   Answer:

$$\frac{40\ \cancel{\text{mg}}}{} \left| \frac{2\ \text{mL}}{125\ \cancel{\text{mg}}} \right| \frac{40 \times 2}{125} \left| \frac{80}{125} \right. = 0.64 \text{ or } 0.6\ \text{mL}$$

CHAPTER 5

Two-Factor Medication Problems

INTRODUCTION

This chapter assists the learner to use dimensional analysis to accurately calculate medication problems involving the weight of the patient, the reconstitution of medications from powder to liquid form, and the amount of time over which medications or intravenous (IV) fluids can be safely administered. To be able to calculate two-factor–given quantity to one-factor or two-factor–wanted quantity medication problems, the learner must understand all factors that may need consideration in medication problems.

Although medications are ordered by physicians and administered by nurses using the five rights of medication administration, other factors might need to be considered when administering certain medications or IV fluids. The weight of the patient often must be factored into a medication problem when determining how much medication can safely be given to an infant, a child, or a geriatric patient. The dosage of medication available may be in powdered form and need reconstitution to a liquid form before parenteral or IV administration. Also, the length of time over which medications or IV fluids can be given plays an important role in the safe administration of IV therapy.

■ Objectives

After completing this chapter, the learner will be able to

1. Solve two-factor–given quantity to one-factor–wanted quantity medication problems involving a specific amount of medication ordered based on the weight of the patient.
2. Calculate medication problems requiring reconstitution of medications by using information from a nursing drug reference, label, or package insert.
3. Solve two-factor–given quantity to two-factor–wanted quantity medication problems involving a specific amount of fluid to be delivered over a limited time using an IV pump delivering milliliters per hour (mL/hr).
4. Solve two-factor–given quantity to two-factor–wanted quantity medication problems involving a specific amount of fluid to be delivered over a limited time using different types of IV tubing that deliver drops per minute (gtt/min) based on a specific drop factor.

■ Teaching Tips

MEDICATION PROBLEMS INVOLVING WEIGHT

Using the five steps involved in problem solving with dimensional analysis, either the sequential method or the random method can be used to calculate two-factor–given quantity medication problems without difficulty. The given quantity (the physician's order) now contains two parts including a numerator (dosage of medication) and a denominator (the weight of the patient). This type of medication problem is called a two-factor medication problem because the given quantity contains two parts (a numerator and a denominator) instead of just one part (a numerator).

It is helpful to make sure that the learner understands that the four terms and the five steps involved in solving problems with dimensional analysis have not changed. Understanding the difference between kilograms and pounds is the most important concept to assist the learner with conceptualizing medication problems involving weight. Most errors occur because students do not correctly factor into the unit path the conversion for weight (example: 1 kg = 2.2 lb). The error in the unit path can easily be identified when checking the conversion factors (error example: 1 lb = 2.2 kg). Student participation in solving the problems, peer teaching, and hands-on experience with medication bottles, labels, or package inserts will promote understanding of the information.

MEDICATION PROBLEMS INVOLVING RECONSTITUTION

Some medications in vials are in a powder form and need reconstitution before administration. Reconstitution involves adding a specific amount of sterile solution (also called diluent) to the vial to change the powder to a liquid form. Information as to how much diluent to add to the vial and what dosage of medication per milliliter will result after reconstitution (also called yield) can be obtained from a nursing drug reference, label, or package insert.

Most students have difficulty conceptualizing reconstitution without hands-on experience; therefore, it is necessary to have vials of medications in powder form, vials of medications in liquid form, and vials of normal saline or sterile water available to practice reconstitution. It also is important that students look up the information from a nursing drug reference, medication label, or package insert to understand the steps that must be followed with reconstitution of a medication.

MEDICATION PROBLEMS INVOLVING INTRAVENOUS PUMPS

Intravenous medications are administered by drawing a specific amount of medication from a vial or ampule and inserting that medication into an existing IV line. All IV medications must be given with specific regard to exactly how much time it should take to administer the medication. Information regarding time may be obtained from a nursing drug reference, medication label, or package insert or may be specifically ordered by the physician.

Although IV medications can be administered IV push, often the time involved requires the use of an intravenous pump (IV pump). All IV pumps deliver milliliters per hour (mL/hr or cc/hr) but may vary in operational capacity or size. It is important for students to understand that when time and the use of an IV pump become involved in the medication order, the given quantity is now a two-factor–given quantity and the wanted quantity is now a two-factor–wanted quantity. Helping the student to understand the concept of medication administration involving IV pumps can be achieved by having an IV pump available for hands-on demonstration and practice. It is necessary for students to see exactly how the answer to the problem (mL/hr or cc/hr) comes into play by actually setting the rate and volume to be infused for any IV medication.

MEDICATION PROBLEMS INVOLVING DROP FACTORS

Although IV pumps are used whenever possible, there are situations (no IV pumps available) and circumstances (outpatient or home care) that arise when IV pumps are not available and IV fluids or medications might be administered using gravity flow. Gravity flow involves calculating the drops per minute (gtt/min) required to infuse IV fluids or medications. When IV fluids or medications are administered using gravity flow, it is important to know the drop factor for the IV tubing that is being used. Drop factor is the drops per milliliter (gtt/mL) that the IV tubing will produce. There are two types of IV tubing available for gravity flow. Macrotubing delivers a large drop and is available in 10 gtt/mL, 15 gtt/mL, and 20 gtt/mL, and microtubing delivers a small drop and is available in 60 gtt/mL. Regardless of the IV tubing used, the five steps can be used and the problem can be solved using dimensional analysis as a problem-solving method.

Hands-on experience is once again necessary to aid the learner with conceptualization through visualization and practice. IV bags of various sizes, IV poles, and a variety of tubing should be available for students to use for practice. Allowing students to see how the IV tubing is packaged and practice identifying the drop factor listed on the box or package will assist with conceptualization. It also is necessary for students to actually count the number of drops per minute based on the answer to the medication problem.

MEDICATION PROBLEMS INVOLVING INTERMITTENT INFUSION

IV medications can be delivered over a specific amount of time by intermittent infusion. When medications are delivered by intermittent infusion, they require the use of an infusion pump. Some medications must be reconstituted and further diluted in a specific type and amount of IV fluid and delivered over a limited time. Other medications do not need to be reconstituted but must be further diluted in a specific type and amount of IV fluid and delivered over a limited time.

This type of medication problem requires two steps and is often difficult for students to understand. It is necessary that visualization be achieved by having equipment available. Students must first be assisted with understanding that the medication may need reconstitution (this information is available from a nursing drug reference, medication label, or package insert) and then further diluted before intravenous piggyback (IVPB) administration. It is important to assist the student to understand that whatever amount is drawn from the medication vial is then inserted into a IV bag (50, 100, or 250 cc of D5W or NS) and that total amount then becomes the amount to be infused. This is best achieved with hands-on experience.

Student Learning Activities for Major Concepts

CONCEPT 1. MEDICATION PROBLEMS INVOLVING WEIGHT (OT 24 THROUGH 29)

- Discuss with the learners how two-factor medication orders from physicians now contain two parts including a numerator (the dosage of medication ordered) and a denominator (the weight of the patient or the time required for safe administration).
- Discuss with the learners how the four terms and the five problem-solving steps used with dimensional analysis continue to apply with two-factor medication problems.
- Discuss with the learners the option of using the sequential method or the random method of dimensional analysis when solving medication problems involving weight.
- Discuss with the learners the common equivalents used when solving medication problems involving weight (2.2 lb = 1 kg, 1 kg = 1000 g, 1 lb = 16 oz).
- Demonstrate for the learners, using Example 5.1, the five steps of problem solving with dimensional analysis using OT 24 through OT 29.
- Set up a learning station for Example 5.1 with the medication bottle, syringe, physician order form, and a scale.

CONCEPT 2. MEDICATION PROBLEMS INVOLVING RECONSTITUTION (OT 30 AND 31)

- Discuss with the learners the definitions of *reconstitution, diluent,* and *yield.*
- Discuss with the learners where to obtain information regarding reconstitution of a medication.
- Demonstrate examples of information regarding reconstitution using overhead transparencies from a nursing drug reference, medication label, or package insert.
- Set up learning stations for Examples 5.2, 5.3, and 5.4 with medication bottles, syringes, sterile water vials, physician order forms, a nursing drug reference, medication labels, and package inserts. Use OT 30 and OT 31 For Example 5.2.

CONCEPT 3. MEDICATION PROBLEMS INVOLVING INTRAVENOUS PUMPS (OT 32 AND 33)

- Discuss with the learners the different methods used to administer IV medications (IV push, intermittent infusion, or continuous infusion).
- Discuss with the learners how time plays an important part in the administration of IV medications and can be obtained from a nursing drug reference, medication label, package insert, or from a physician as part of the medication order.
- Discuss with the learners the different types of IV pumps and delivery methods (mL/hr or cc/hr).
- Demonstrate for the learners, using Example 5.5, the five steps of problem solving with dimensional analysis using OT 32 and OT 33.
- Set up learning stations for Examples 5.5, 5.6, and 5.7 with IV bags, physician order forms, and IV pumps.
- Demonstrate for the learners how information can be obtained by the nurse from the IV pump to determine if the dosage of medication the patient is receiving is within a safe dosage range using Example 5.8 and the overhead transparencies for page 112.

CONCEPT 4. MEDICATION PROBLEMS INVOLVING DROP FACTORS

- Discuss with the learners the definitions of *gravity flow* and *drop factor.*
- Explore with the learners when situations might arise when IV pumps are not available.
- Discuss with the learners, using Table 5.1, the different types of IV tubing available for administration of IV fluids using gravity flow.
- Demonstrate for the learners, using Example 5.9, the five steps of problem solving with dimensional analysis.
- Demonstrate for the learners, using Example 5.10, the five steps of problem solving with dimensional analysis.
- Discuss with the learners how the conversion factor 1 hr = 60 min is used in most medication problems calculating gtt/min.
- Demonstrate for the learners, using Example 5.11, how the nurse can calculate the number of hours it will take for a specific amount of IV fluid to infuse.
- Explore with the learners the importance of checking an IV infusion every 2 hours to prevent air in the line when an IV totally infuses.
- Explore with the learners why the answer for Example 5.11 is not rounded up.
- Set up learning stations for Examples 5.9, 5.10, and 5.11 with IV bags, physician order forms, IV pole, and various packaged IV tubing.

CONCEPT 5. MEDICATION PROBLEMS INVOLVING INTERMITTENT INFUSION

- Explore with the learners the different types of IV intermittent infusion pumps that can be used to deliver medication over a specific time (secondary IV pumps and mini-infusers).
- Demonstrate for the learners, using Example 5.12, the two steps required for calculating this medication with dimensional analysis.
- Set up a learning station for Example 5.12 with the medication vial, a vial of sterile water, physician order form, syringes, IV bag, IV pump, IV pole, 10 gtt/mL IV tubing, and a nursing drug reference.

■ Frequently Asked Questions

The most frequently asked questions regarding the information in Chapter 5: Two-Factor Medication Problems include:

1. Why is the weight of the patient important in a medication order?

   Answer: The weight of the patient often must be factored into a medication problem when determining how much medication can safely be given to an infant, child, or geriatric patient. The dosage of medication ordered often depends on the weight of the patient to ensure therapeutic levels.
2. Why is there no denominator when factoring the weight of the patient into the unit path?

   Answer: The denominator for the weight of the patient is actually the patient (in Example 5.1, 60 lb = patient) and has no effect on the unit path. Therefore, it can be omitted.
3. How do you know what diluent to use to reconstitute medications?

   Answer: A nursing drug reference, medication label, or package insert will provide information on how much and what type of diluent to use to reconstitute a drug as well as what would be the yield or reconstitution.
4. How do you know how much time is needed to infuse an IV medication?

   Answer: Information regarding the specific amount of time to safely administer an IV medication can be obtained from a nursing drug reference, medication label, or package insert, or may be specifically ordered by the physician.
5. How do you know which method to use to administer IV medications?

   Answer: Although IV medications can be administered IV push, often the time involved requires the use of an IV pump. Information regarding the specific method that can be used to safely administer an IV medication can be obtained from a nursing drug reference, medication label, or package insert.
6. What types of IV tubing are available to use for gravity flow?

   Answer: There are two types of IV tubing available for gravity flow. Macrotubing delivers a large drop and is available in 10 gtt/mL, 15 gtt/mL, and 20 gtt/mL; and microtubing delivers a small drop and is available in 60 gtt/mL.
7. How can you prevent an IV from running dry?

   Answer: It is safe nursing practice to monitor an infusing IV every 2 hours to make sure it is infusing without difficulty and on time. The nurse must be prepared to hang the IV at least 30 minutes before the next IV bag is due to be hung.
8. How do you decide what type of IV fluid to use to further dilute a reconstituted drug?

   Answer: A nursing drug reference provides information regarding what type of IV fluid to use for further dilution and the amount of time needed to infuse the IV fluid.

■ Conclusion

This chapter assisted the learner to calculate two-factor medication problems involving the weight of the patient, reconstitution of medications, and the amount of time over which medications and IV fluids can be safely administered using the sequential method or the random method of dimensional analysis. The learner should have a clear understanding of the concepts in Chapter 5 before proceeding to Chapter 6 and three-factor medication problems.

POST-TEST FOR CHAPTER 5: TWO-FACTOR MEDICATION PROBLEMS

Name ______________________________ Date ______________

1. Order: Furosemide 2 mg/kg PO every 8 hours for congestive heart failure.

   The child weighs 10 kg.

   How many milliliters will you give?

   Answer: Sequential method

$$\frac{2\ \cancel{\text{mg}}}{\cancel{\text{kg}}} \;\Big|\; \frac{\cancel{10\ \text{kg}}}{} \;\Big|\; \frac{\text{mL}}{\cancel{10\ \text{mg}}} = 2\ \text{mL}$$

2. Order: Meperidine 1.5 mg/kg PO every 4 hours for pain. The child weighs 22 lb. How many milliliters will you give?

   Answer: Random method

$$\frac{1.5\ \cancel{\text{mg}}}{\cancel{\text{kg}}} \;\Big|\; \frac{5\ \text{mL}}{50\ \cancel{\text{mg}}} \;\Big|\; \frac{1\ \cancel{\text{kg}}}{2.2\ \cancel{\text{lb}}} \;\Big|\; \frac{22\ \cancel{\text{lb}}}{} \;\Big|\; \frac{1.5 \times 5 \times 1 \times 22}{50 \times 2.2} \;\Big|\; \frac{165}{110} = 1.5\ \text{mL}$$

3. Order: Epogen 100 units/kg IV tid for anemia secondary to chronic renal failure. The patient weighs 160 lb. How many milliliters will you give?

   Answer: Sequential method

$$\frac{\cancel{100}\ \cancel{\text{units}}}{\cancel{\text{kg}}} \;\Big|\; \frac{1\ \cancel{\text{kg}}}{2.2\ \cancel{\text{lb}}} \;\Big|\; \frac{16\cancel{0}\ \cancel{\text{lg}}}{} \;\Big|\; \frac{\text{mL}}{\cancel{4000}\ \cancel{\text{units}}} \;\Big|\; \frac{1 \times 1 \times 16}{2.2 \times 4} \;\Big|\; \frac{16}{8.8} = 1.81 \text{ or } 1.8\ \text{mL}$$

4. Order: Augmentin 10 mg/kg PO every 8 hours for otitis media. Nursing drug reference states: Dilute with one teaspoon (5 mL) of tap water and shake vigorously to yield 125 mg per 5 mL. The child weighs 25 kg. How many milliliters will you give after reconstitution?

   Answer: Random method

$$\frac{10\ \cancel{\text{mg}}}{\cancel{\text{kg}}} \;\Big|\; \frac{5\ \text{mL}}{125\ \cancel{\text{mg}}} \;\Big|\; \frac{25\ \cancel{\text{kg}}}{} \;\Big|\; \frac{10 \times 5 \times 25}{125} \;\Big|\; \frac{1250}{125} = 10\ \text{mL}$$

(Post test continues on page 24)

5. The physician orders heparin to infuse at 1300 units/hr continuous IV infusion. The pharmacy sends an IV bag labeled heparin 25,000 units in 250 mL. Calculate the milliliters per hour to set the IV pump.

   Answer: Sequential method

$$\frac{1300\ \text{units}}{\text{hr}}\ \Big|\ \frac{250\ \text{mL}}{25{,}000\ \text{units}}\ \Big|\ 13 = \frac{13\ \text{mL}}{\text{hr}}$$

6. A patient is receiving heparin 25,000 units in 250 mL infusing at 25 mL/hr. How many units per hour is the patient receiving?

   Answer: Sequential method

$$\frac{25\ \text{mL}}{\text{hr}}\ \Big|\ \frac{25{,}000\ \text{units}}{250\ \text{mL}}\ \Big|\ \frac{25 \times 2500}{25}\ \Big|\ \frac{62500}{25} = \frac{2500\ \text{units}}{\text{hr}}$$

7. The physician orders morphine sulfate 2 mg/hr continuous IV for intractable pain related to end-stage lung cancer. The pharmacy sends an IV bag labeled morphine sulfate 100 mg in 250 mL. Calculate milliliters per hour to set the IV pump.

   Answer: Sequential method

$$\frac{2\ \text{mg}}{\text{hr}}\ \Big|\ \frac{250\ \text{mL}}{100\ \text{mg}}\ \Big|\ \frac{2 \times 25}{10}\ \Big|\ \frac{55}{10} = \frac{5\ \text{mL}}{\text{hr}}$$

8. Order: 1000 mL D5W/1/2 NS with 20 mEq of KCl to infuse in 12 hours.

   Drop factor: 20 gtt/mL

   Calculate the number of drops per minutes.

   Answer: Sequential method

$$\frac{1000\ \text{mL}}{12\ \text{hr}}\ \Big|\ \frac{20\ \text{gtt}}{\text{mL}}\ \Big|\ \frac{1\ \text{hr}}{60\ \text{min}}\ \Big|\ \frac{1000 \times 2 \times 1}{12 \times 6}\ \Big|\ \frac{2000}{72} = 27.77 \text{ or } \frac{28\ \text{gtt}}{\text{min}}$$

9. Order: Azactam 500 mg IV every 12 hours for septicemia.

Supply: Azactam 1-g vials

Nursing drug reference states: Dilute each 1-g vial with 10 mL of sterile water for injection and further dilute in 100 mL of NS to infuse over 60 minutes.

How many milliliters will you draw from the vial after reconstitution?

Calculate milliliters per hour to set the IV pump.

Calculate drops per minute with a drop factor of 20 gtt/mL.

Answer: Random method

$$\frac{5\cancel{00}\ \cancel{mg}}{} \bigg| \frac{1\cancel{0}\ mL}{1\ \cancel{g}} \bigg| \frac{1\ \cancel{g}}{1\cancel{000}\ \cancel{mg}} \bigg| \frac{5 \times 1 \times 1}{1 \times 1} \bigg| \frac{5}{1} = 5\ mL$$

$$\frac{105\ mL}{\cancel{60\ min}} \bigg| \frac{\cancel{60\ min}}{1\ hr} \bigg| \frac{105}{1} = \frac{105\ mL}{hr}$$

$$\frac{105\ \cancel{mL}}{\cancel{hr}} \bigg| \frac{2\cancel{0}\ gtt}{\cancel{mL}} \bigg| \frac{1\ \cancel{hr}}{6\cancel{0}\ min} \bigg| \frac{105 \times 2 \times 1}{6} \bigg| \frac{210}{6} = \frac{35\ gtt}{min}$$

10. Order: Ancef 6.25 mg/kg IV every 6 hours for pneumonia. The child weighs 38.2 kg.

Nursing drug reference states: Dilute each 1-g vial with 10 mL of sterile water for injection and further dilute in 50 mL of NS to infuse over 30 minutes.

How many milliliters will you draw from the vial after reconstitution?

Calculate milliliters per hour to set the IV pump.

Calculate drops per minute with a drop factor of 10 gtt/mL.

Answer: Random method

$$\frac{6.25\ \cancel{mg}}{\cancel{kg}} \bigg| \frac{1\cancel{0}\ mL}{1\ \cancel{g}} \bigg| \frac{1\ \cancel{g}}{100\cancel{0}\ \cancel{mg}} \bigg| \frac{38.2\ \cancel{kg}}{} \bigg| \frac{6.25 \times 1 \times 1 \times 38.2}{1 \times 100} \bigg| \frac{238.75}{100}$$

$$= 2.3875\ mL$$

$$\text{or } 2.4\ mL$$

$$\frac{52.4\ mL}{3\cancel{0}\ \cancel{min}} \bigg| \frac{6\cancel{0}\ \cancel{min}}{1\ hr} \bigg| \frac{52.4 \times 6}{3 \times 1} \bigg| \frac{314.4}{3} = \frac{104.8\ mL}{hr}$$

$$\frac{104.8\ \cancel{mL}}{\cancel{hr}} \bigg| \frac{1\cancel{0}\ gtt}{\cancel{mL}} \bigg| \frac{\cancel{1}\ \cancel{hr}}{6\cancel{0}\ min} \bigg| \frac{104.8 \times 1}{6} \bigg| \frac{104.8}{6} = 17.466 \text{ or } \frac{17\ gtt}{min}$$

CHAPTER 6

Three-Factor Medication Problems

INTRODUCTION

This chapter assists the learner to accurately calculate medication problems involving the dosage of medication based on the weight of the patient and the time required for safe administration using dimensional analysis as a problem-solving method. When medications are ordered by physicians for infants, children, or the elderly, the dosage of medication (g, mg, mcg, gr) based on the weight of the patient must be considered as well as how much medication the patient can receive per dose or day. Although the physician orders the medications, the nurse must be aware of the safe dosage range for administration of medications to infants and children as well as adults.

When medications are ordered by physicians for critically ill patients, the patients must be closely monitored by the nurse for effectiveness of the medications. Often, the medications or intravenous (IV) fluids must be titrated for effectiveness, with an increase or decrease in the dosage based on the patient's response to the medications. Factors involved in the safe administration of medications or IV fluids for the critically ill patient include the dosage of medication based on the combined factors of the weight of the patient and the time required for administration. The medication may need reconstitution or preparation by the nurse for immediate administration in a critical situation. The weight of the patient also may need to be obtained daily to ensure accurate correlation with the dosage of medication ordered.

To be able to calculate three-factor–given quantity to one-factor–, two-factor–, or three-factor–wanted quantity medication problems, the learner must understand all of the components of the medication order and be able to calculate medication problems in a critical situation.

■ Objectives

After completing this chapter, the learner will be able to

1. Calculate three-factor–given quantity to one-factor–, two-factor–, or three-factor–wanted quantity medication problems involving a specific amount of medication or IV fluid based on the weight of the patient and the time required for safe administration.
2. Calculate problems requiring reconstitution or preparation of medications using information from a nursing drug reference, label, or package insert.

■ Teaching Tips

MEDICATION PROBLEMS INVOLVING DOSAGE, WEIGHT, AND TIME

Using the five problem-solving steps, three-factor–given quantity medication problems can be solved implementing the sequential method or the random method of dimensional analysis. The given quantity or the physician's order now contains three parts including a numerator (the dosage of medication ordered) and two denominators (the weight of the patient and the time required for safe administration).

The most difficult concept for the learner to understand is that there are now two denominators but that this does not change the way the problem is solved using dimensional analysis. The correct answer is still achieved by canceling out all unwanted units from the unit path either using the sequential or random method of dimensional analysis. When implementing the sequential or the random method, the medication problem can be set up in a number of different ways with a focus on the correct placement of conversion factors to allow unwanted units to be canceled from the unit path. Errors occur because the weight of the patient is often factored into the unit path incorrectly, the conversion for weights is reversed, or the arithmetic is incorrect.

Student participation in solving the problems, peer teaching, and hands-on experience with all of the required equipment assist the student to understand all the necessary steps. Learning stations are one of the best teaching/learning strategies to use to ensure student competence. Learning stations provide students the opportunity to calculate the problem and then to actually implement the answer with hands-on experience. The learning station should contain all the necessary supplies and equipment for the student to use (IV pumps or IV set-up with pole, medication drawer with syringes, medication cups, and various sizes and types of IVPB bags) and several different types of medication calculation problems.

■ Student Learning Activities for Major Concepts

CONCEPT 1. MEDICATION PROBLEMS INVOLVING DOSAGE, WEIGHT, AND TIME (OT 34 THROUGH 45)

- Discuss with the learners how three-factor medication orders from physicians now contain three parts including a numerator (the dosage of medication ordered) and two denominators (the weight of the patient and the time required for safe administration).
- Discuss with the learners how the four terms and five problem-solving steps used with dimensional analysis continue to apply with three-factor medication problems.
- Discuss with the learners the option of using the sequential method or the random method of dimensional analysis when solving medication problems involving three factors.
- Demonstrate for the learner, using Example 6.1, the five steps of problem solving with dimensional analysis using OT 34 through OT 41.
- Set up a learning station for Example 6.1 with the medication bottle, medication syringe, physician order form, and a scale.
- Demonstrate for the learners how information can be obtained by the nurse from the medication sheet to determine if a patient is receiving a dosage of medication within a safe dosage range using Example 6.2 and OT 42 through OT 45.
- Explore with the learners how the nursing drug reference identifies a safe dosage range for medication administration.
- Demonstrate for the learner, using Example 6.3, the five steps of problem solving with dimensional analysis.
- Discuss with the learners the concept of titrating medication for effectiveness.
- Demonstrate for the learners how information can be obtained by the nurse from the IV pump to determine if a patient is receiving a dosage of medication within a safe dosage range using Example 6.4.
- Set up a learning station for Example 6.3 with an IV bag, IV pump, and a physician's order form.

■ Frequently Asked Questions

The most frequently asked questions regarding the information in Chapter 6: Three-Factor Medication Problems include

1. Why does the nurse have to know safe dosage ranges if the physician orders the medications?

   Answer: The nurse is part of the health care team and shares the legal obligation to ensure that the patient receives medication within the safe dosage range of all medications administered.

2. How often do you have to titrate medications?

   Answer: When medications are ordered by physicians for critically ill patients, the patients must be closely monitored by the nurse for effectiveness of medications based on the hemodynamic readings for the patient. Hemodynamic readings are monitored frequently by the nurses based on the time frequencies ordered by the physician (every 30 minutes, every 1 to 2 hours, or every 4 hours).

■ Conclusion

Dimensional analysis is a problem-solving method that nurses can use to calculate a variety of medication problems in the hospital, outpatient, or the home care environment. This chapter has assisted the learner to calculate three-factor medication problems involving the dosage of medication, the weight of the patient, and the amount of time over which medications or IV fluids can be safely administered.

With advanced nursing and home care nursing resulting in increased autonomy, it is more important than ever that nurses be able to accurately calculate medication problems. Dimensional analysis provides the opportunity to use one problem-solving method for any type of medication problem, thereby increasing consistency and decreasing confusion when calculating medication problems.

POST-TEST FOR CHAPTER 6: THREE-FACTOR MEDICATION PROBLEMS

Name ______________________________ Date ______________

1. Order: morphine sulfate 0.3 mg/kg/dose PO every 4 hours for pain.

   Supply: morphine sulfate 10 mg/5 mL.

   Child's weight: 20 lb

   How many milliliters per dose will you give?

   Answer: Random method

$$\begin{array}{c|c|c|c} 0.3\ \cancel{mg} & 5\ \text{mL} & 2\cancel{0}\ \cancel{lb} & 1\ \cancel{kg} \\ \hline \cancel{kg}\ /\ \text{dose} & 1\cancel{0}\ \cancel{mg} & & 2.2\ \cancel{lb} \end{array} = \frac{\text{mL}}{\text{dose}}$$

$$\begin{array}{c|c} 0.3 \times 5 \times 2 \times 1 & 3 \\ \hline 1 \times 2.2 & 2.2 \end{array} = 1.36 \text{ or } \frac{1.4\ \text{mL}}{\text{dose}}$$

2. Order: filgrastim 5 mcg/kg/day for myelosuppression secondary to chemotherapy administration.

   Supply: filgrastim 480 mcg/1.6 mL

   Patient's weight: 100 lb

   How many milliliters per day will you give?

   Answer: Sequential method

$$\begin{array}{c|c|c|c} 5\ \cancel{mcg} & 1.6\ \text{mL} & 1\ \cancel{kg} & 10\cancel{0}\ \cancel{lb} \\ \hline \cancel{kg}\ /\ \text{day} & 48\cancel{0}\ \cancel{mcg} & 2.2\ \cancel{lb} & \end{array} = \frac{\text{mL}}{\text{day}}$$

$$\begin{array}{c|c} 5 \times 1.6 \times 1 \times 10 & 80 \\ \hline 48 \times 2.2 & 105.6 \end{array} = \frac{0.757}{\text{day}} \text{ or } 0.76 \text{ or } 0.8\ \text{mL}$$

(Post test continues on page 30)

3. Order: Epogen 100 units/kg/day SQ three times weekly for anemia secondary to AZT administration.

   Supply: Epogen 10,000 units/mL

   Patient's weight: 180 lb

   How many milliliters per day will you give?

   Answer: Sequential method

$$\frac{1\cancel{00}\ \cancel{\text{units}}}{\cancel{\text{kg}}\ /\ \text{day}} \;\Big|\; \frac{\text{mL}}{10{,}\cancel{000}\ \cancel{\text{units}}} \;\Big|\; \frac{1\ \cancel{\text{kg}}}{2.2\ \cancel{\text{lb}}} \;\Big|\; \frac{18\cancel{0}\ \cancel{\text{lb}}}{} = \frac{\text{mL}}{\text{day}}$$

$$\frac{1 \times 1 \times 18}{10 \times 2.2} \;\Big|\; \frac{18}{22} = 0.818 \text{ or } 0.82 \text{ or } \frac{0.8\ \text{mL}}{\text{day}}$$

4. Order: digoxin 25 mcg/kg/day PO every 8 hours for congestive heart failure.

   Supply: digoxin 0.25 mg per 5 mL

   Child's weight: 25 lb

   How many milliliters will you give per day?

   How many milliliters will you give per dose?

   Answer: Random method

$$\frac{25\ \cancel{\text{mcg}}}{\cancel{\text{kg}}\ /\ \text{day}} \;\Big|\; \frac{5\ \text{mL}}{0.25\ \cancel{\text{mg}}} \;\Big|\; \frac{25\ \cancel{\text{lb}}}{} \;\Big|\; \frac{1\ \cancel{\text{mg}}}{1000\ \cancel{\text{mcg}}} \;\Big|\; \frac{1\ \cancel{\text{kg}}}{2.2\ \cancel{\text{lb}}} = \frac{\text{mL}}{\text{day}}$$

$$\frac{25 \times 5 \times 25 \times 1 \times 1}{0.25 \times 1000 \times 2.2} \;\Big|\; \frac{3125}{550} = 5.68 \text{ or } 5.7 \text{ or } \frac{6\ \text{mL}}{\text{day}}$$

   Answer: Random method

$$\frac{25\ \cancel{\text{mcg}}}{\cancel{\text{kg}}\ /\ \cancel{\text{day}}} \;\Big|\; \frac{5\ \text{mL}}{0.25\ \cancel{\text{mg}}} \;\Big|\; \frac{25\ \cancel{\text{lb}}}{} \;\Big|\; \frac{1\ \cancel{\text{mg}}}{1000\ \cancel{\text{mcg}}} \;\Big|\; \frac{1\ \cancel{\text{kg}}}{2.2\ \cancel{\text{lb}}} \;\Big|\; \frac{\cancel{\text{day}}}{3\ \text{dose}} = \frac{\text{mL}}{\text{dose}}$$

$$\frac{25 \times 5 \times 25 \times 1 \times 1}{0.25 \times 1000 \times 2.2 \times 3} \;\Big|\; \frac{3125}{1650} = 1.89 \text{ or } 1.9 \text{ or } \frac{2\ \text{mL}}{\text{dose}}$$

5. Order: clindamycin 10 mg/kg/day IV in three divided doses for respiratory tract infection.

   Supply: clindamycin 150 mg/mL

   Child's weight: 10 kg

   How many milliliters per dose will you draw from the vial?

   Answer: Sequential method

10 ~~mg~~	1~~0~~ ~~kg~~	~~day~~	(mL)	10 × 1	10
~~kg~~ / ~~day~~		3 (doses)	15~~0~~ ~~mg~~	3 × 15	45

$= 0.22 \text{ or } \frac{0.2 \text{ mL}}{\text{dose}}$

6. Order: Claforan 100 mg/kg/day IV in two divided doses for infection

   Supply: Claforan 1 gm/10 mL

   Neonate's weight: 2045 gm

   How many milliliters per dose will you draw from the vial?

   Answer: Random method

1~~00~~ ~~mg~~	1 ~~kg~~	2045 ~~gm~~	1~~0~~ (mL)	1 ~~gm~~	~~day~~
~~kg~~ / ~~day~~	10~~00~~ ~~gm~~		1 ~~gm~~	10~~00~~ ~~mg~~	2 (doses)

$= \frac{\text{mL}}{\text{dose}}$

1 × 1 × 2045 × 1 × 1	2045
10 × 1 × 100 × 2	2000

$= 1.022 \text{ or } \frac{1 \text{ mL}}{\text{dose}}$

7. Order: gentamicin 2.5 mg/kg/dose IV every 12 hours for gram-negative bacillary infection.

   Supply: gentamicin 40 mg/mL

   Neonate's weight: 1182 g

   How many milligrams per dose will the neonate receive?

   Answer: Random method

2.5 (mg)	1182 ~~g~~	1 ~~kg~~	2.5 × 1182 × 1	2955
~~kg~~ / (dose)		1000 ~~g~~	1000	1000

$= 2.955 \text{ or } \frac{3.0 \text{ mg}}{\text{dose}}$

(Post test continues on page 32)

8. Order: ampicillin 100 mg/kg/day IV in divided doses every 12 hours for respiratory tract infection.

   Supply: ampicillin 125 mg/5 mL

   Neonate's weight: 1182 g

   How many milligrams per dose will the neonate receive?

   How many milliliters per dose will you draw from the vial?

   Answer: Random method

$$\frac{1\cancel{00}\ \text{mg}}{\cancel{\text{kg}}/\cancel{\text{day}}}\ \Big|\ \frac{1182\ \cancel{\text{g}}}{}\ \Big|\ \frac{1\ \cancel{\text{kg}}}{10\cancel{00}\ \cancel{\text{g}}}\ \Big|\ \frac{\cancel{\text{day}}}{2\ \text{doses}}\ \Big|\ \frac{1 \times 1182 \times 1}{10 \times 2}\ \Big|\ \frac{1182}{20} = \frac{59.1\ \text{mg}}{\text{dose}}$$

$$\frac{59.1\ \cancel{\text{mg}}}{\text{dose}}\ \Big|\ \frac{5\ \text{mL}}{125\ \cancel{\text{mg}}}\ \Big|\ \frac{59.1 \times 5}{125}\ \Big|\ \frac{295.5}{125} = 2.364 \text{ or } 2.36 \text{ or } \frac{2.4\ \text{mL}}{\text{dose}}$$

9. Order: Solu-Medrol 5.4 mg/kg/hr IV for acute spinal cord injury.

   Supply: Solu-Medrol 125 mg/2 mL

   Patient's weight: 160 lb

   How many milligrams per hour will the patient receive?

   Answer: Sequential method

$$\frac{5.4\ \text{mg}}{\cancel{\text{kg}}/\text{hr}}\ \Big|\ \frac{1\ \cancel{\text{kg}}}{2.2\ \cancel{\text{lb}}}\ \Big|\ \frac{160\ \cancel{\text{lb}}}{}\ \Big|\ \frac{5.4 \times 1 \times 160}{2.2}\ \Big|\ \frac{864}{2.2} = 392.72 \text{ or } \frac{392.7\ \text{mg}}{\text{hr}}$$

10. Order: aminophylline 0.8 mg/kg/hr IV for respiratory distress.

    Supply: aminophylline 250 mg/100 mL

    Child's weight: 65 lb

    Calculate milliliters per hours to set the IV pump.

    Answer: Sequential method

$$\frac{0.8\ \cancel{\text{mg}}}{\cancel{\text{kg}}/\text{hr}}\ \Big|\ \frac{10\cancel{0}\ \text{mL}}{25\cancel{0}\ \cancel{\text{mg}}}\ \Big|\ \frac{1\ \cancel{\text{kg}}}{2.2\ \cancel{\text{lb}}}\ \Big|\ \frac{65\ \cancel{\text{lb}}}{} = \frac{\text{mL}}{\text{hr}}$$

$$\frac{0.8 \times 10 \times 1 \times 65}{25 \times 2.2}\ \Big|\ \frac{520}{55} = 9.45 \text{ or } \frac{9.5\ \text{mL}}{\text{hr}}$$

SECTION 2

Practice Problems

INTRODUCTION

This section assists the learner to practice calculating one-factor, two-factor, and three-factor medication problems using dimensional analysis. The goal of this section is to clarify that the learner has a clear understanding of dimensional analysis as a problem-solving method and is capable of solving a variety of medication problems accurately.

■ Teaching Tips

This section is an independent learning unit. Each learner should work through the medication problems in this section with a goal of 100% accuracy.

■ Conclusion

This section has provided the learner with the opportunity for extensive practice with a variety of medication problems using dimensional analysis as a problem-solving method. This section also has provided the learner the opportunity to review one-factor, two-factor, and three-factor medication problems to evaluate comprehension.

SECTION 3

Case Studies

INTRODUCTION

This section contains 20 case studies that will assist learners to practice reading orders written by the physician to simulate an actual clinical nursing setting. The case studies contain a variety of one-factor, two-factor, and three-factor medication calculation problems. The goal of this section is to provide learners with an opportunity to experience reading and interpreting orders written by the physician using dimensional analysis as a problem-solving method.

■ Teaching Tips

Simulating an actual clinical nursing setting through the use of learning stations is the best teaching/learning strategy for the learner. Allowing the student to interpret and implement orders from a physician provides a means of evaluating if the student has acquired the mathematical, conceptual, and cognitive skills necessary to provide safe patient care.

■ Conclusion

This section provides learners with the opportunity to evaluate mathematical, conceptual, and cognitive skills through simulation of an actual clinical nursing setting using dimensional analysis as a problem-solving method. Regardless of whether it is used in acute care, long-term care, or home care, dimensional analysis remains an avenue for problem solving when the goal is improvement of medication dosage calculation abilities, reduction of medication errors, and patient safety.

SECTION 4

Comprehensive Post-Test

INTRODUCTION

This section includes 20 comprehensive medication calculation problems to allow learners the opportunity to evaluate their ability to calculate medication problems using dimensional analysis. A score of 100%, or all 20 problems answered correctly, is an ideal maximum goal for the learner. A score of 80%, or 16 problems answered correctly, is a minimum goal. A score of 100% demonstrates competence and ensures patient safety when calculating a variety of medication problems using dimensional analysis as a problem-solving method.

■ Teaching Tips

This section is a post-test that allows learners to assess their knowledge level and competence. The learner should be encouraged to use a calculator but should show computation for each problem to allow for evaluation of problem-solving skills. The learner should also be allowed to refer to the conversion table to identify conversion factors necessary to solve the problems.

■ Conclusion

The 20 medication problems chosen for the post-test represent the types of medication calculation problems that a registered nurse would encounter with medication administration in clinical nursing practice. The reliability of the testing instrument was determined through use of the odd–even split half-test of reliability (0.714).

COMPREHENSIVE POST-TEST

Name ______________________________ Date ______________

1. Order: Phenobarbital 60 mg PO daily for seizures

   Supply on hand: Phenobarbital 30 mg/tablet

   How many tablets will you give?

   Answer: Sequential method

6~~0~~ ~~mg~~	(tablets)	6
	3~~0~~ ~~mg~~	3

= 2 tablets

2. Order: Chloral hydrate 250 mg PO 30 minutes before hs as sedative

   Supply on hand: Chloral hydrate 250 mg/5 mL

   How many milliliters will you give?

   Answer: Sequential method

~~250 mg~~	5 (mL)	5
	~~250 mg~~	

= 5 mL

3. Order: Digitoxin 0.3 mg PO daily for maintenance dose following digitalization

   Supply on hand: Digitoxin 100 mcg/tablet

   How many tablets will you give?

   Answer: Random method

0.3 ~~mg~~	(tablets)	10~~00~~ ~~mcg~~	0.3 × 10	3
	1~~00~~ ~~mcg~~	1 ~~mg~~	1 × 1	1

= 3 tablets

4. Order: Potassium chloride 20 mEq PO tid for hypokalemia

   Supply: Potassium chloride 40 mEq/15 mL

   How many teaspoons will you give?

   Answer: Sequential method

2~~0~~ ~~mEq~~	15 ~~mL~~	1 (tsp)	2 × 15 × 1	30
	4~~0~~ ~~mEq~~	5 ~~mL~~	4 × 5	20

= 1.5 tsp

(Post test continues on page 40)

5. Order: 500 mL D5W to infuse over 12 hours

   Drop factor: 60 gtt/mL

   Calculate the number of drops per minute.

   Answer: Sequential method

$$\frac{500\ \cancel{mL}}{12\ \cancel{hr}} \bigg| \frac{\cancel{60}\ gtt}{\cancel{mL}} \bigg| \frac{1\ \cancel{hr}}{\cancel{60}\ min} \bigg| \frac{500 \times 1}{12} \bigg| \frac{500}{12} = 41.66 \text{ or } \frac{42\ gtt}{min}$$

6. Order: Heparin 1500 units/hr for thrombophlebitis

   Supply: Heparin 25,000 units/500 cc

   Calculate cc/hr to set the IV pump.

   Answer: Sequential method

$$\frac{15\cancel{00}\ \cancel{units}}{hr} \bigg| \frac{50\cancel{0}\ cc}{25{,}\cancel{000}\ \cancel{units}} \bigg| \frac{15 \times 50}{25} \bigg| \frac{750}{25} = \frac{30\ cc}{hr}$$

7. Order: Infuse heparin at 20 cc/hr for thrombophlebitis

   Supply: Heparin 25,000 units in 250 cc

   How many units/hr is the patient receiving?

   Answer: Sequential method

$$\frac{2\cancel{0}\ \cancel{cc}}{hr} \bigg| \frac{25{,}000\ units}{25\cancel{0}\ \cancel{cc}} \bigg| \frac{2 \times 25000}{25} \bigg| \frac{50000}{25} = \frac{2000\ units}{hr}$$

8. Order: Infuse bolus of 0.9% NS at 100 gtt/min

   Supply: 250 mL 0.9% NS with 60 gtt/mL tubing

   How many hours will it take to infuse the IV bolus?

   Answer: Sequential method

$$\frac{25\cancel{0}\ \cancel{mL}}{} \bigg| \frac{\cancel{60}\ \cancel{gtt}}{\cancel{mL}} \bigg| \frac{\cancel{min}}{10\cancel{0}\ \cancel{gtt}} \bigg| \frac{1\ hr}{\cancel{60}\ \cancel{min}} \bigg| \frac{25 \times 1}{10} \bigg| \frac{25}{10} = 2.5 \text{ hours}$$

9. Order: Fluconazole 200 mg IVPB over 60 minutes for systemic candidal infections

Supply: Fluconazole 200 mg/100 mL with 20 gtt/mL tubing

Calculate the number of drops per minute.

Answer: Sequential method

100 ~~mL~~	2~~0~~ (gtt)	100 × 2	200	= 33.3 or $\frac{33\ gtt}{min}$
6~~0~~ (min)	~~mL~~	6	6	

10. Order: Furosemide 2 mg/kg PO daily for congestive heart failure

Supply: Furosemide 10 mg/mL oral solution

How many milliliters will you give a child weighing 10 lb?

Answer: Sequential method

2 ~~mg~~	(mL)	1 ~~kg~~	~~10 lb~~	2 × 1	2	= 0.9 mL
~~kg~~	~~10 mg~~	2.2 ~~lb~~		2.2	2.2	

11. Order: Neupogen 6 mcg/kg SQ twice daily for chronic neutropenia

Supply: Neupogen 300 mcg/mL

How many milliliters will you give a client weighing 175 lb?

Answer: Sequential method

6 ~~mcg~~	(mL)	1 ~~kg~~	175 ~~lb~~	6 × 1 × 175	1050	= 1.59
~~kg~~	300 ~~mcg~~	2.2 ~~lb~~		300 × 2.2	660	or 1.6 mL

(Post test continues on page 42)

12. Order: Ampicillin 500 mg IV every 6 hours for urinary tract infection

Supply: Ampicillin 1-g vial

Nursing drug reference: Reconstitute each 1-g vial with 10 mL of sterile water and further dilute in 50 mL of 0.9% NS and infuse over 15 minutes.

How many milliliters will you draw from the vial after reconstitution?

Calculate the milliliters per hour to set the IV pump.

Calculate the drops per minute with a drop factor of 10 gtt/mL.

Answer: Sequential method

$$\frac{5\cancel{00}\ \cancel{\text{mg}}}{} \left| \frac{1\cancel{0}\ \text{mL}}{\cancel{1\ \text{g}}} \right| \frac{\cancel{1\ \text{g}}}{1\cancel{000}\ \cancel{\text{mg}}} \left| \frac{5 \times 1}{1} \right| \frac{5}{1} = 5\ \text{mL}$$

$$\frac{55\ \text{mL}}{15\ \cancel{\text{min}}} \left| \frac{60\ \cancel{\text{min}}}{1\ \text{hr}} \right| \frac{55 \times 60}{15 \times 1} \left| \frac{3300}{15} \right. = \frac{220\ \text{mL}}{\text{hr}}$$

$$\frac{220\ \cancel{\text{mL}}}{\cancel{\text{hr}}} \left| \frac{1\cancel{0}\ \text{gtt}}{\cancel{\text{mL}}} \right| \frac{1\ \cancel{\text{hr}}}{6\cancel{0}\ \text{min}} \left| \frac{220 \times 1 \times 1}{6} \right| \frac{220}{6} = 36.66 \text{ or } \frac{37\ \text{gtt}}{\text{min}}$$

13. Order: Acyclovir 10 mg/kg IV every 8 hours for varicella zoster in immunosuppressed patient weighing 140 lb

Supply: Acyclovir 1-g vial

Nursing drug reference: Reconstitute each 1-g vial with 10 mL of sterile water and further dilute in 100 mL of 0.9% NS and infuse over 1 hour.

How many milliliters will you draw from the vial after reconstitution?

Calculate milliliters per hour to set the IV pump.

Calculate the drop per minute with a drop factor of 10 gtt/mL.

Answer: Sequential method

$$\frac{1\cancel{0}\ \cancel{\text{mg}}}{\cancel{\text{kg}}} \left| \frac{1\cancel{0}\ \text{mL}}{\cancel{1\ \text{g}}} \right| \frac{\cancel{1\ \text{g}}}{1\cancel{000}\ \cancel{\text{mg}}} \left| \frac{1\ \cancel{\text{kg}}}{2.2\ \cancel{\text{lb}}} \right| \frac{14\cancel{0}\ \cancel{\text{lb}}}{} \left| \frac{1 \times 1 \times 1 \times 14}{1 \times 2.2} \right| \frac{14}{2.2} = 6.36 \text{ or } 6.4\ \text{mL}$$

$$\frac{106.4\ \text{mL}}{1\ \text{hr}} \left| \frac{106.4}{1} \right. = \frac{106\ \text{mL}}{\text{hr}}$$

$$\frac{106\ \cancel{\text{mL}}}{\cancel{\text{hr}}} \left| \frac{1\cancel{0}\ \text{gtt}}{\cancel{\text{mL}}} \right| \frac{1\ \cancel{\text{hr}}}{6\cancel{0}\ \text{min}} \left| \frac{106 \times 1 \times 1}{6} \right| \frac{106}{6} = 17.66 \text{ or } \frac{18\ \text{gtt}}{\text{min}}$$

14. Order: Epinephrine 1 mcg/min IV for bradycardia

    Supply: Epinephrine 1 mg/250 mL 0.9% NS

    Calculate milliliters per hour to set the IV pump.

    Answer: Random method

~~1 mcg~~	25~~0~~ mL	~~1 mg~~	6~~0~~ ~~min~~	25 × 6	150	=	15 mL
~~min~~	~~1 mg~~	10~~00~~ ~~mcg~~	~~1~~ hr	10	10		hr

15. Order: Isuprel 5 mcg/min IV for heart block

    Supply: Isuprel 2 mg in 500 mL D5W

    Calculate milliliters per hour to set the IV pump.

    Answer: Random method

5 ~~mcg~~	5~~00~~ mL	~~1 mg~~	6~~0~~ ~~min~~	5 × 5 × 6	150	=	75 mL
~~min~~	2 ~~mg~~	1~~000~~ ~~mcg~~	~~1~~ hr	2 × 1	2		hr

16. Order: Dobutamine 2.5 mcg/kg/min IV for management of heart failure for a patient weighing 130 lb

    Supply: Dobutamine 250 mg in 1000 mL of 0.9% NS

    Calculate milliliters per hour to set the IV pump.

    Answer: Random method

2.5 ~~mcg~~	~~1000~~ mL	1 ~~mg~~	1 ~~kg~~	6~~0~~ ~~min~~	130 ~~lb~~	=	mL
~~kg/min~~	25~~0~~ ~~mg~~	1~~000 mcg~~	2.2 ~~lb~~	1 hr			hr

2.5 × 1 × 1 × 6 × 130	1950	= 35.45 or	35.5 mL
25 × 2.2 × 1	55		hr

17. Order: Dopamine 10 mcg/kg/min IV for management of hypotension secondary to decreased cardiac output for a patient weighing 120 lb

    Supply: Dopamine 400 mg in 500 mL D5W

    Calculate milliliters per hour to set the IV pump.

    Answer: Random method

1~~0~~ ~~mcg~~	5~~00~~ mL	~~1 mg~~	6~~0~~ ~~min~~	~~1 kg~~	12~~0 lb~~	=	mL
~~kg/min~~	4~~00 mg~~	~~1000 mcg~~	~~1~~ hr	2.2 ~~lb~~			hr

1 × 5 × 6 × 12	360	= 40.9 or	41 mL
4 × 2.2	8.8		hr

(Post test continues on page 44)

18. Order: Aminophylline is infusing at 24 mL/hr for respiratory distress for a patient weighing 80 kg

Supply: Aminophylline is 250 mg in 250 mL in D5W

How many mg/kg/hr is the patient receiving?

Answer: Sequential method

$$\frac{24\ \cancel{mL}}{hr}\ \bigg|\ \frac{\cancel{250}\ mg}{\cancel{250\ mL}}\ \bigg|\ \frac{}{80\ kg}\ \bigg|\ \frac{24}{80} = \frac{0.3\ mg}{kg/hr}$$

19. Order: Amrinone infusing at 47 mL/hr for a patient weighing 100 kg for short-term treatment of congestive heart failure

Supply: Amrinone 100 mg/100 mL 0.45% NS

How many mcg/kg/min is the patient receiving?

Answer: Sequential method

$$\frac{47\ \cancel{mL}}{\cancel{hr}}\ \bigg|\ \frac{\cancel{100\ mg}}{\cancel{100\ mL}}\ \bigg|\ \frac{\cancel{1000}\ mcg}{\cancel{1\ mg}}\ \bigg|\ \frac{}{\cancel{100}\ kg}\ \bigg|\ \frac{\cancel{1\ hr}}{6\cancel{0}\ min} = \frac{mcg}{kg/min}$$

$$\frac{47}{6} = 7.8 \text{ or } \frac{8\ mcg}{kg/min}$$

20. Order: Aminophylline loading dose of 5.6 mg/kg to infuse over 30 minutes for a patient weighing 50 kg followed by 0.6 mg/kg/hr maintenance dose for COPD

Supply: Aminophylline 500 mg in 500 mL of D5W

How many milliliters per hour will you set the IV pump for the loading dose?

How many milliliters per hour will you set the IV pump for the maintenance dose?

Answer: Sequential method

$$\frac{5.6\ \cancel{mg}}{\cancel{kg}/3\cancel{0}\ \cancel{min}}\ \bigg|\ \frac{\cancel{500}\ mL}{\cancel{500\ mg}}\ \bigg|\ \frac{5\cancel{0}\ \cancel{kg}}{}\ \bigg|\ \frac{60\ \cancel{min}}{1\ hr}\ \bigg|\ \frac{5.6 \times 5 \times 60}{3 \times 1}\ \bigg|\ \frac{1680}{3} = \frac{560\ mL}{hr}$$

$$\frac{0.6\ \cancel{mg}}{\cancel{kg}/hr}\ \bigg|\ \frac{\cancel{500}\ mL}{\cancel{500\ mg}}\ \bigg|\ \frac{50\ \cancel{kg}}{}\ \bigg|\ \frac{0.6 \times 50}{}\ \bigg|\ \frac{30}{} = \frac{30\ mL}{hr}$$

Dimensional Analysis

#1

Four Basic Terms Used with Dimensional Analysis:

- Given Quantity (the beginning point of the problem)
- Wanted Quantity (the answer to the problem)
- Unit Path (the series of conversions necessary to achieve the answer to the problem)
- Conversion Factors (equivalents necessary to convert between systems of measurement and to allow unwanted units to be canceled from the problem)

Dimensional Analysis

The FIVE LEARNING PRINCIPLES:

- **Identify the GIVEN QUANTITY of the problem**
- **Identify the WANTED QUANTITY of the problem**
- **Establish the UNIT PATH from the GIVEN QUANTITY to the WANTED QUANTITY using equivalents as CONVERSION FACTORS**
- **Set up the problem to permit cancellation of unwanted units**
- **Multiply NUMERATORS, Multiply DENOMINATORS, and Divide the product of the numerators by the product of the denominators to provide the numerical value of the WANTED QUANTITY**

Dimensional Analysis

#3

UNIT PATH GRID:

Given Quantity	Conversion Factor for Given Quantity	Conversion Factor for Wanted Quantity	Conversion Computation	=	Wanted Quantity

Dimensional Analysis

#4

Problem: 1 liter (L) equals how many ounces (oz)?

Given Quantity =

Wanted Quantity =

Conversions =

Dimensional Analysis

#5

Problem: 1 Liter = How many ounces?

1. Identify the GIVEN QUANTITY in the problem.

Example: Given Quantity = 1 Liter (L)

Given Quantity

1 Liter (L)		=

Dimensional Analysis

2. Identify the WANTED QUANTITY in the problem.

Example: Wanted Quantity = Ounces (oz)

Given Quantity			Wanted Quantity
1 Liter (L)		=	oz

Dimensional Analysis

#7

3. Establish the UNIT PATH from the GIVEN QUANTITY to the WANTED QUANTITY factoring in the CONVERSION FACTORS.

Given Quantity	Conversion Factors			Wanted Quantity
1 Liter (L)	1000 ml	1 oz	=	oz
	1 L	30 ml		

Dimensional Analysis

#8

4. Write the setup for the problem so that unwanted units are canceled from the UNIT PATH and wanted units are correctly placed to correlate with the WANTED QUANTITY.

Given Quantity	Conversion Factors			Wanted Quantity
1 ~~L~~	1000 ~~ml~~	1 oz	=	oz
	1 ~~L~~	30 ~~ml~~		

Dimensional Analysis

#9

5. Multiply the numerators, multiply the denominators, and divide the product of the numerators by the product of the denominators to provide the numerical value of the WANTED QUANTITY.

Given Quantity	Conversion Factors		Computation		Wanted Quantity
1 L	1000 ml	1 oz	1000	=	33.3 oz
	1 L	30 ml	30		

Dimensional Analysis

#10

Problem: 1 gallon (gal) = How many milliliters (ml) or How many ml in 1 gal?

Given Quantity =

Wanted Quantity =

Conversion =

Dimensional Analysis

Problem: 1 gallon (gal) equals how many milliliter (ml)?

1. Identify the GIVEN QUANTITY in the problem.

Example: Given Quantity = 1 gallon (gal)

Given Quantity

$$\frac{1 \text{ gallon (gal)}}{\quad} \Bigg| \frac{\qquad\qquad}{\qquad\qquad} =$$

Dimensional Analysis

2. Identify the WANTED QUANTITY in the problem.

Example: Wanted Quantity = ml

Given Quantity		Wanted Quantity
1 gal		= ml

Dimensional Analysis

#13

3. Establish the UNIT PATH from the GIVEN QUANTITY to the WANTED QUANTITY factoring in the CONVERSION FACTORS

Given Quantity	Conversion Factors				Wanted Quantity
1 gal	4 qt	1 L	1000 ml	=	ml
	1 gal	1 qt	1 L		

Dimensional Analysis

#14

4. Write the setup for the problem so that unwanted units are canceled from the UNIT PATH and wanted units are correctly placed to correlate with the WANTED QUANTITY.

Given Quantity	Conversion Factors			Wanted Quantity
1 gal	4 qt	1 L	1000 ml	= ml
	1 gal	1 qt	1 L	

Dimensional Analysis

5. **Multiply the numerators, multiply the denominators, and divide the product of the numerators by the product of the denominators to provide the numerical value of the WANTED QUANTITY.**

Given Quantity	Conversion Factors			Computation		Wanted Quantity
1 gal	4 qt	1 L	1000 ml	4 x 1000	=	4000 ml
	1 gal	1 qt	1 L	1		

Dimensional Analysis

One-Factor Medication Problems

By using the Five Steps of Dimensional Analysis, all medication problems can be solved with the:

- **SEQUENTIAL METHOD** (canceling unwanted units from the preceding factor)
- **RANDOM METHOD** (canceling unwanted units without regard for the preceding factor but correlating correct placement of CONVERSION FACTORS with the WANTED QUANTITY)

Dimensional Analysis

PROBLEM EXAMPLE #1

The physician orders gr 10 aspirin orally every 4 hours as needed for fever. The unit dose of medication on hand is gr 5 per tablet (5 gr/tab). How many tablets will you administer?

SEQUENTIAL METHOD:

Given Quantity =

Wanted Quantity =

Dose on Hand =

Dimensional Analysis

Step #1. Identify the Given Quantity (the physician's order)

Given Quantity

$$\frac{10 \text{ gr}}{\quad} \Bigg| \frac{\quad}{\quad} =$$

Dimensional Analysis

Step #2. Identify the Wanted Quantity (the answer to the problem)

Given Quantity — **Wanted Quantity**

10 gr | ______ = **tablets**

Dimensional Analysis

Step #3. Establish the Unit Path from the Given Quantity to the Wanted Quantity using Equivalents as Conversion Factors.

Given Quantity	Conversion Factors		Wanted Quantity
10 gr	tablets	=	tablets
	5 gr		

$$\frac{10\ \text{gr}}{} \times \frac{\text{tablets}}{5\ \text{gr}} = \text{tablets}$$

Dimensional Analysis

#21

Step #4. Set up the problem to allow cancellation of unwanted units.

Given Quantity	Conversion Factors	Wanted Quantity
$\frac{10\ \cancel{gr}}{}$	$\frac{\text{tablets}}{5\ \cancel{gr}}$	= tablets

Dimensional Analysis

#22

Step #5. Multiply numerators, multiply denominators, and divide to provide the numerical value for the Wanted Quantity.

Given Quantity	Conversion Factors	Computation	Wanted Quantity
10 gr	tablets / 5 gr	10 / 5	= 2 tablets

$$\frac{10\ \cancel{gr}}{} \left| \frac{\text{tablets}}{5\ \cancel{gr}} \right| \frac{10}{5} = 2 \text{ tablets}$$

Dimensional Analysis

#23

The FIVE LEARNING PRINCIPLES are still applicable with TWO-FACTOR medication problems but now:

- The GIVEN QUANTITY has a NUMERATOR and a DENOMINATOR
- The WANTED QUANTITY may have a NUMERATOR and a DENOMINATOR

Dimensional Analysis

Medication Problems Involving Weight

PROBLEM EXAMPLE #1

The physician order Gentamicin 2.5 mg/kg IV every 8 hours for infection. The vial of medication is labeled 40 mg/ml. The child weighs 60 lb. How many ml will you give?

Given Quantity =

Wanted Quantity =

Dose on Hand =

Weight =

Dimensional Analysis

Step #1. Identify the Given Quantity
SEQUENTIAL METHOD:

Given Quantity

$$\frac{2.5\text{ mg}}{\text{kg}} \Bigg| \underline{\qquad\qquad\qquad} =$$

Dimensional Analysis

#26

Step #2. Identify the Wanted Quantity

SEQUENTIAL METHOD:

Given Quantity		Wanted Quantity
$\frac{2.5\ \text{mg}}{\text{kg}}$		= ml

Dimensional Analysis

#27

Step #3. Establish the Unit Path

SEQUENTIAL METHOD:

Given Quantity	Conversion Factors		Wanted Quantity
$\frac{2.5\ mg}{kg}$	$\frac{ml}{40\ mg}$	=	ml

Dimensional Analysis

Step #4. Cancel unwanted units
SEQUENTIAL METHOD:

Given Quantity	Conversion Factors			Wanted Quantity
2.5 ~~mg~~	ml	1 ~~kg~~	60 ~~lb~~	= ml
~~kg~~	40 ~~mg~~	2.2 ~~lb~~		

Dimensional Analysis

#29

Step #5. Multiply and Divide
SEQUENTIAL METHOD:

Given Quantity		Conversion Factors		Computation			Wanted Quantity
2.5 ~~mg~~	ml	1 ~~kg~~	60 ~~lb~~	2.5 x 1 x 6	15	=	1.7 ml
~~kg~~	40 ~~mg~~	2.2 ~~lb~~		4.22	8.8		

Dimensional Analysis

Medication Problems Involving Reconstitution

PROBLEM EXAMPLE #2

The physician orders Mezlin 50 mg/kg every 4 hours IV for infection. The child weighs 60 lb. Pharmacy sends a vial labeled Mezlin 1 g. NDR: Reconstitute 1 g with 10 ml of sterile water. How many ml will you draw from the vial?

Given Quantity =

Wanted Quantity =

Dose on Hand =

Weight =

Dimensional Analysis

#31

RANDOM METHOD:

Given Quantity	Conversion Factors				Wanted Quantity
50 ~~mg~~	1 ~~kg~~	60 ~~lb~~	10 ml	1 ~~gm~~	= ml
~~kg~~	2.2 ~~lb~~		1 ~~gm~~	1000 ~~mg~~	

Computation

5 x 6	30
2.2	2.2

= 13.63 or 13.6 ml is the answer

Dimensional Analysis

Medication Problems Involving Intravenous Pumps

PROBLEM EXAMPLE #5

The physician orders Heparin 1500 Units/hr IV. The supply is Heparin 25,000 Units/250 ml D5W. Calculate ml/hr to set the IV pump.

Given Quantity =

Wanted Quantity =

Dose on Hand =

Dimensional Analysis

SEQUENTIAL METHOD:

Given Quantity	Conversion Factors	Computation		Wanted Quantity
1500 ~~Units~~	250 ml	15	=	15 ml
hr	25,000 ~~Units~~			hr

Dimensional Analysis

Medication Problems Involving Dosage, Weight, and Time

PROBLEM EXAMPLE #1

The physician orders Tagamet for GI ulcers, 30 mg/kg/day PO in four divided doses for a child weighing 22 kg. The dose on hand is Tagamet 300 mg/5 ml. How many ml per day will the child receive?

Given Quantity =

Wanted Quantity =

Dose on Hand =

Weight =

Dimensional Analysis

#35

SEQUENTIAL METHOD:

Given Quantity			Wanted Quantity
30 mg / kg/day		=	ml / day

$$\frac{30\ \text{mg}}{\text{kg/day}} \Bigg| \qquad = \frac{\text{ml}}{\text{day}}$$

Dimensional Analysis

SEQUENTIAL METHOD:

Given Quantity	Conversion Factor		Wanted Quantity
30 mg / kg/day	5 ml / 300 mg	=	ml / day

$$\frac{30\ \text{mg}}{\text{kg/day}} \times \frac{5\ \text{ml}}{300\ \text{mg}} = \frac{\text{ml}}{\text{day}}$$

Dimensional Analysis

#37

SEQUENTIAL METHOD:

Given Quantity	Conversion Factors			Wanted Quantity
30 mg	5 ml	22 kg	=	ml
kg/day	300 mg			day

Dimensional Analysis

SEQUENTIAL METHOD:

Given Quantity	Conversion Factors		Computation			Wanted Quantity
30 mg	5 ml	22 kg	3 x 5 x 22	330	=	ml
kg/day	300 mg		30	30		day

$$\frac{330}{30} = \frac{11 \text{ ml}}{\text{day}}$$ in four divided doses

Dimensional Analysis

#39

How many milliliters per dose will the child receive?

Given Quantity =

Wanted Quantity =

Dimensional Analysis

#40

SEQUENTIAL METHOD:

11 ml	
day	

= ml / dose

Dimensional Analysis

#41

SEQUENTIAL METHOD:

$$\frac{11\text{ ml}}{\text{day}} \,\Big|\, \frac{\text{day}}{4\text{ doses}} \,\Big|\, \frac{11}{4} = \frac{2.75\text{ or }2.8\text{ ml}}{\text{dose}}$$

Dimensional Analysis

PROBLEM EXAMPLE #2

As a prudent nurse, you are concerned that the child may be receiving an unsafe dosage of Tagamet. The child weighs 22 kg and is receiving 2.8 ml/dose QID. The dose on hand is 300 mg/5 ml. Calculate mg/kg/day.

Given Quantity =

Wanted Quantity =

Dose on Hand =

Weight =

Dimensional Analysis

#43

SEQUENTIAL METHOD:

$$\frac{2.8\ \text{ml}}{\text{dose}} \Bigg| \frac{\quad}{\quad} = \frac{\text{mg}}{\text{kg/day}}$$

Dimensional Analysis

#44

SEQUENTIAL METHOD:

$$\frac{2.8\ \cancel{ml}}{\cancel{dose}} \left| \frac{300\ mg}{5\ \cancel{ml}} \right| \frac{4\ \cancel{doses}}{day} \left| \frac{}{22\ kg} \right. = \frac{mg}{kg/day}$$